Abdelhafid Mimouni

Oxidative stress in Streptococcus buccale

Abdelhafid Mimouni

Oxidative stress in Streptococcus buccale

ScienciaScripts

Cover image: www.ingimage.com

This book is a translation from the original published under ISBN 978-613-8-43123-7.

Publisher:
Sciencia Scripts
is a trademark of
Dodo Books Indian Ocean Ltd. and OmniScriptum S.R.L publishing group

120 High Road, East Finchley, London, N2 9ED, United Kingdom
Str. Armeneasca 28/1, office 1, Chisinau MD-2012, Republic of Moldova, Europe
Printed at: see last page
ISBN: 978-620-8-26321-8

Oxidative stress in Streptococcus buccale

Author : Dr. Abdelhafid Mimouni

An independent researcher in bioinorganic chemistry, Dr Mimouni is an expert in macromolecular synthesis and characterisation. He obtained his PhD in Chemistry from the University of Paris XII in 1997, after a Diplôme des Études Approfondies in Bioinorganic Systems from the University of Paris XI in 1993, where he also obtained his Licence and Maîtrise in Chemistry.

Summary: This study explores oxidative stress in *Streptococcus buccale*, highlighting its role in pathogenesis and toxin production. Oxidative stress, generated by internal and external factors, influences the virulence of this bacterium. Metalloenzymes such as superoxide dismutase and catalase act as antioxidant defences, protecting the bacteria from reactive oxygen species. The toxins produced, particularly exotoxins, affect the host cells, with potentially neurological implications. The accumulation of these toxins in the brain could have adverse effects on cognitive health. The interaction between oxidative stress and toxin production highlights the importance of future research to develop targeted treatments. The results also indicate significant links with various oral and neurological diseases, reinforcing the importance of an integrated approach to health.

Book outline:

Introduction

Oxidative stress is a biological condition characterised by an imbalance between the production of reactive oxygen species (ROS) and the body's ability to neutralise their harmful effects through antioxidant mechanisms. ROS, which include free radicals and other oxidising molecules, play a crucial role in various biological processes, but their excessive accumulation can lead to damage to cells, proteins and DNA. This phenomenon is recognised as a contributing factor in numerous pathologies, such as cardiovascular disease, diabetes and neurodegenerative disorders.

In the field of oral microbiology**, Streptococcus buccale** stands out for its important role. Present in the microbial flora of the oral cavity, this bacterium contributes to the balance of the oral ecosystem, while also being associated with pathologies such as dental caries and periodontal infections. What makes it particularly relevant is its ability to generate and respond to oxidative stress.

The study of oxidative stress mechanisms in **Streptococcus buccale** is essential for understanding how this bacterium interacts with its environment, modifies its virulence in response to stress, and influences oral and systemic health. The aim of this book is to explore the effects of oxidative stress on **oral Streptococcus** and the implications for health.

Chapter 1: Introduction to Streptococcus buccale

Morphology and classification

Streptococcus buccale is a Gram-positive bacterium belonging to the Streptococcaceae family. Its morphology is characterised by a spherical or oval shape, and it often groups together in chains or pairs. The size of these bacteria generally varies between 0.5 and 1.0 µm in diameter. As a species of the genus Streptococcus**, S. buccale** shares characteristics with other streptococci, but is distinguished by its specific adaptation to the oral environment.

Streptococci are classified according to their haemolysis on a culture medium. Although **S. buccale** is mainly non-haemolytic, its precise classification can be influenced by biochemical and molecular tests, including genetic analysis. This allows a better understanding of its role and interactions with other bacterial species present in the oral cavity.

Habitat and role in the oral cavity

Streptococcus buccale is widely present in the oral cavity, where it colonises tooth surfaces, mucous membranes and gums. As a member of the normal microbial flora, this bacterium plays an essential role in maintaining the oral balance. It participates in the initial digestion of carbohydrates by breaking down complex sugars, which contributes to the host's nutrition. In addition, **S. buccale** helps to protect against other pathogens by producing antimicrobial substances and occupying ecological niches, thereby limiting the colonisation of pathogens.

However, under certain conditions, such as inadequate oral hygiene, an imbalance in the microbial flora can encourage the proliferation **of Streptococcus buccale**. When this bacterium multiplies excessively, it can produce acids by fermenting the sugars present in the mouth. These acids can then demineralise tooth enamel, leading to tooth decay and other oral infections such as gingivitis and periodontitis.

Thus, although **S. buccale** is a normal component of the oral flora, its role in health and disease depends on the delicate balance of the oral ecosystem. A thorough understanding of these dynamics is essential for developing strategies to prevent and treat oral diseases.

Chapter 2: Mechanisms of oxidative stress

Sources of oxidative stress

Oxidative stress in **Streptococcus buccale** comes from a number of sources, both internal and external. Internal metabolic processes, including cellular respiration, generate reactive oxygen species (ROS), including superoxides and peroxides. These ROS are normal by-products of cellular metabolism, but their excessive accumulation can damage biomolecules, including lipids, proteins and DNA.

In addition, environmental factors can exacerbate ROS production. Exposure to pathogens, such as other bacteria or viruses, can induce an inflammatory response that increases ROS production. Similarly, exposure to ultraviolet light or harmful chemicals (such as antibiotics or disinfectants) can also cause oxidative stress. This stress can alter cellular function and influence the virulence **of oral Streptococcus**, underlining the importance of understanding these sources in the context of oral health.

Detection mechanisms and stress response

Streptococcus buccale has sophisticated mechanisms for detecting and responding to oxidative stress. These bacteria use intracellular signalling systems that enable them to assess ROS levels. For example, specific regulatory proteins can detect changes in ROS concentrations and activate signalling pathways that trigger adaptive responses.

In response to increased stress, **Streptococcus buccale** activates specific genes that code for defence proteins. Among these, superoxide dismutase (SOD) plays a key role, converting superoxides into hydrogen peroxide, a less harmful compound. Other antioxidant enzymes are also mobilised, enabling the bacteria to strengthen their defences against oxidative stress.

Role of metalloenzymes

Metalloenzymes, in particular superoxide dismutase (SOD), are crucial to the defence of **Streptococcus buccale** against oxidative stress. SOD catalyses the conversion of superoxides to hydrogen peroxide, providing a first line of defence. Although some strains of **Streptococcus** do not possess catalase, the enzyme that breaks down hydrogen peroxide into water and oxygen, they may have other enzyme systems to neutralise this compound. For example, peroxidases and glutathione peroxidase (GPx) may be present, playing a complementary role in the degradation of ROS.

GPx, in particular, is important because it uses glutathione to reduce hydrogen peroxide, thereby protecting the cell from oxidative damage. The presence and activity of these antioxidant enzymes are essential for the survival and virulence **of Streptococcus buccale** in an oral environment often rich in oxidative stress.

Conclusion of the chapter

This chapter highlights the defence mechanisms of **Streptococcus buccale** against oxidative stress, underlining the importance of these processes in the survival and virulence of this bacterium in the oral cavity. The ability to detect and respond effectively to oxidative stress is crucial for maintaining equilibrium in the oral ecosystem. A better understanding of these mechanisms may pave the way for therapeutic strategies aimed at mitigating the harmful effects of oxidative stress on human health, particularly with regard to oral infections and associated diseases.

Chapter 3: Metalloenzymes and antioxidant systems

Role of superoxide dismutase (SOD)

Superoxide dismutase (SOD) is an essential metalloenzyme present in **Streptococcus buccale**. It catalyses the dismutation of superoxides (O_2-) into hydrogen peroxide (H_2O_2) and oxygen (O_2). This process is vitally important, as superoxides are particularly reactive molecules that can cause significant damage to cell membranes, proteins and DNA. By reducing superoxide levels, SOD plays a crucial protective role in maintaining cellular integrity and bacterial survival.

SOD activity is regulated according to the levels of ROS present in the environment. When an increase in superoxide levels is detected, **Streptococcus buccale** activates the genes associated with SOD, ensuring a rapid and effective response to oxidative stress. This defence mechanism is fundamental to the bacterium's resistance to the unfavourable environmental conditions encountered in the oral cavity.

Protection mechanisms against reactive oxygen species (ROS)

In addition to SOD**, Streptococcus buccale** deploys several antioxidant systems to protect itself from ROS:

1. **Catalase**: Although some strains of **Streptococcus** do not possess it, catalase is a key enzyme that breaks down hydrogen peroxide into water and oxygen. This process contributes significantly to

protection against oxidative damage, by reducing the toxicity of hydrogen peroxide, an intermediate product generated by SOD.

2. **Glutathione peroxidase (GPx)**: This enzyme uses glutathione, a powerful antioxidant, to reduce hydrogen peroxide. GPx thus offers additional protection against oxidative stress by neutralising peroxides, minimising the risk of cell damage.
3. **Metallothioneins**: These small proteins with the ability to bind metal ions play an important role in detoxification and protection against oxidative stress. They bind to heavy metals and free radicals, reducing the harmful effects of these elements in the cell.

These antioxidant systems do not function in isolation; they are part of a complex network of cellular responses aimed at maintaining homeostasis. Their interaction strengthens the resilience of **Streptococcus buccale** to oxidative stress.

Interactions with other metabolic pathways

The antioxidant systems of **Streptococcus buccale** also interact with other metabolic pathways. For example, the regulation of iron metabolism is crucial, as excessive levels of iron can promote the production of ROS via the Fenton reaction, where iron catalyses the conversion of hydrogen peroxide into hydroxyl radicals, even more reactive species. This highlights the importance for bacteria to balance their acquisition and storage of iron while maintaining effective defence mechanisms.

In addition, the metabolism of glutathione, a key antioxidant, is closely linked to amino acid synthesis pathways and the stress response. The use of glutathione precursors can be modulated according to the needs of the cell, allowing rapid adaptation to variations in oxidative stress.

Conclusion

This chapter explored the central role of metalloenzymes and antioxidant systems in the defence of **Streptococcus buccale** against oxidative stress. Understanding these mechanisms highlights the importance of antioxidant enzymes in bacterial virulence and survival. Furthermore, this knowledge may offer potential avenues for the development of targeted therapies aimed at alleviating oral infections, by targeting bacterial protective pathways and improving oral health.

Chapter 4: Toxins produced by Streptococcus

Types of toxins

Streptococcus buccale produces several types of toxin which play a crucial role in its virulence and interaction with the host. These toxins include :

1. **Exotoxins**: These proteins secreted **by Streptococcus buccale** have various cytotoxic effects on host cells. They include enzymes such as streptolysins, which are capable of lysing red blood cells and other cell types, causing tissue damage. These exotoxins can also modulate the immune response, facilitating evasion of host defence mechanisms.
2. **Membrane toxins**: Certain strains **of Streptococcus buccale** are capable of producing toxins that disrupt cell membranes. These toxins can cause tissue damage, promoting inflammation and infection in the oral cavity. Their action can weaken tissue integrity, which can have significant clinical consequences.

Mechanisms of action and impact on host cells

The toxins produced by **Streptococcus buccale** act through several mechanisms, directly impacting host cells:

1. **Cell lysis**: Streptolysins, for example, bind to the cell membranes of host cells and force the formation of pores. This disruption of

membrane integrity leads to cell death, allowing the bacteria to spread and infect surrounding tissues.

2. **Inflammation**: Toxins can also induce an inflammatory response. By triggering the release of pro-inflammatory cytokines, they exacerbate tissue damage. This immune response can, in turn, foster an environment conducive to bacterial colonisation and survival.
3. **Disruption of cell signalling**: Certain toxins have the ability to interfere with the intracellular signalling pathways of host cells. By disrupting these pathways, they compromise normal cell function, which can lead to cell dysfunction and long-term pathological consequences.

These mechanisms of action underline the pathogenic potential of **Streptococcus buccale** and its ability to cause oral diseases. The bacterium's ability to produce these toxins and manipulate immune responses is a key factor in its virulence.

Conclusion of the chapter

This chapter has highlighted the different types of toxins produced by **Streptococcus buccale** and their mechanisms of action. Exotoxins and membrane toxins play a key role in the pathogenicity of the bacterium, facilitating its invasion and contributing to oral diseases. Understanding these toxins and their impact on the host is essential for developing effective prevention and treatment strategies against oral infections associated with this bacterium.

Chapter 5: Penetration of toxins into the brain

Mechanisms of transport and accumulation of toxins in the nervous system

The ability of toxins produced by **Streptococcus buccale** to penetrate the brain raises major neurological health concerns. The mechanisms involved in this process are complex and deserve particular attention:

1. **Blood transport**: Toxins can enter the bloodstream from sites of oral infection. Once in the blood, they can cross the blood-brain barrier (BBB) by passive diffusion, depending on their size and solubility, or via specific transporters that facilitate their passage. This mechanism is of particular concern, as it allows toxins to reach the central nervous system (CNS), where they can have deleterious effects.
2. **Local inflammation**: The presence of toxins in the oral cavity can trigger an inflammatory response, leading to the release of pro-inflammatory mediators. These substances can alter the permeability of the BBB, facilitating the entry not only of toxins but also of pathogens into the CNS. Chronic inflammation can also contribute to the degradation of BBB integrity, increasing the risk of neuroinvasion.
3. **Neuronal invasion**: Certain strains **of Streptococcus buccale** have the ability to invade nerve cells directly. By penetrating neurons, they can transport their toxins inside these cells, where

they can cause significant damage. This neuronal invasion represents a direct mode of action that further complicates the host's defence mechanisms.

Neurological effects and impact on cognitive health

The accumulation of toxins in the brain can lead to a number of worrying neurological consequences:

1. **Neurotoxicity**: Toxins can cause cellular damage within neurons, leading to cell death and neuronal dysfunction. These effects can manifest as cognitive impairment, memory loss and behavioural changes.
2. **Neurochemical imbalance**: By interfering with the release of neurotransmitters, toxins can influence mood, cognition and other essential neurological functions. This imbalance can result in symptoms such as anxiety, depression or mood disorders.
3. **Neurodegenerative diseases**: Chronic exposure to **Streptococcus buccale** toxins could contribute to the development of neurodegenerative diseases such as Alzheimer's or other forms of dementia. By inducing oxidative stress and inflammation in the brain, these toxins may promote pathological processes associated with neuronal degeneration.

Conclusion

This chapter has explored the mechanisms by which **Streptococcus buccale** toxins can enter the brain and the potential adverse effects on neurological health. Understanding these processes is crucial to establishing links between oral infections and neurological disease. By identifying the pathways by which these toxins interact with the central nervous system, it is possible to better target therapeutic interventions to prevent or mitigate the neurological consequences of oral infections.

Chapter 6: Interactions between oxidative stress and toxin production

Influence of oxidative stress on toxin production

Oxidative stress plays a fundamental role in regulating toxin production by **Streptococcus buccale**. Increased levels of reactive oxygen species (ROS) in the cellular environment can activate several intracellular signalling pathways, leading to increased expression of genes specific to toxin production. This complex interaction between oxidative stress and toxin production can be described as follows:

1. **Activation of the stress response**: Conditions of oxidative stress trigger an adaptive response in **Streptococcus buccale**. Transcription factors such as Nrf2 (Nuclear factor erythroid 2-related factor 2) are activated in response to increased ROS. Nrf2 not only regulates the production of antioxidant proteins that help neutralise oxidative stress, but also influences the expression of genes encoding certain toxins. In this way, oxidative stress can both stimulate defence mechanisms and promote the production of toxins, creating a delicate balance.
2. **Feedback**: The toxins produced by **Streptococcus buccale** can themselves generate additional oxidative stress within the host. For example, by causing cellular damage, these toxins increase the release of ROS, fuelling a vicious circle. This feedback not only amplifies the virulence of the bacteria, but also complicates the

host's immune response, making it more difficult to fight the infection.

The role of toxins in aggressiveness and virulence

The toxins produced by **Streptococcus buccale** are essential to its pathogenicity. They contribute to various aspects of the aggressiveness and virulence of this bacterium, in particular :

1. **Tissue damage**: Toxins have the ability to degrade host tissues. For example, certain exotoxins can lyse cells and cause tissue necrosis. This damage facilitates bacterial invasion, allowing **Streptococcus buccale** to spread and exacerbate inflammation in surrounding tissues. This process can lead to more serious infections and clinical complications.
2. **Immune evasion**: Toxins also play a crucial role in immune evasion. Some of them can inhibit innate and adaptive immune responses, by disrupting immune cell signalling or neutralising cytokines. This allows the bacteria to persist and multiply in the oral cavity, increasing their pathogenic potential.
3. **Interaction with the microbiome**: The production of toxins by **Streptococcus buccale** influences not only its immediate environment, but also the overall composition of the oral microbiome. By favouring a more acidic environment and

eliminating certain competing species, **Streptococcus buccale** creates an environment conducive to its own colonisation and that of other pathogens. This interaction can exacerbate microbial imbalances, increasing the risk of oral infections and systemic complications.

Conclusion

This chapter has examined the complex interactions between oxidative stress and toxin production by **Streptococcus buccale**. Understanding these interactions is essential to better understand how this bacterium contributes to oral diseases and its potential neurological implications. By revealing the mechanisms by which oxidative stress influences virulence and toxin production, we can envisage new therapeutic approaches to prevent and treat infections associated with **oral Streptococcus**.

Chapter 7: Clinical consequences and implications

Link between oxidative stress, toxins and oral and neurological diseases

The link between oxidative stress, the production of toxins by **Streptococcus buccale** and the onset of various diseases is well documented and merits particular attention. This link manifests itself significantly in several oral pathologies, as well as in neurological conditions:

1. **Dental caries**: The formation of dental caries is exacerbated by the degradation of dental tissue caused by the action of toxins. **Streptococcus buccale**, by producing acids during its metabolism, contributes to the demineralisation of tooth enamel. In the presence of oxidative stress, this degradation is accelerated, as ROS promote inflammation and impair the repair capacity of dental tissue. The formation of biofilms, often enriched with this bacterium, also promotes the persistence of cavities.
2. **Gingivitis and periodontitis**: Gum disease, such as gingivitis and periodontitis, is also influenced by oxidative stress and toxins. These trigger intense inflammatory responses, leading to destruction of the tissues supporting the teeth. The chronic inflammation associated with these conditions can be exacerbated by ROS, which impair immune cell function and worsen the inflammatory response, creating a vicious circle.

In terms of neurological implications, oxidative stress and toxins play a crucial role in :

1. **Dementia**: Epidemiological studies show an association between oral infections and the onset of neurodegenerative diseases, such as Alzheimer's disease. Infection-induced oxidative stress can cause neuronal damage and promote the accumulation of toxic proteins such as amyloid, thereby increasing the risk of dementia.
2. **Cognitive disorders**: The accumulation of toxins in the brain has a direct impact on cognitive function. These toxins can disrupt neurotransmitters, cause synaptic alterations and affect neuronal plasticity, thus contributing to cognitive disorders and general cognitive decline.

Therapeutic and preventive approaches

Understanding the mechanisms of interaction between oxidative stress and toxin production opens the way to new therapeutic and preventive approaches:

1. **Antioxidants**: The use of antioxidant compounds could help reduce the deleterious effects of oxidative stress. Studies have shown that the administration of antioxidants can reduce the production of toxins and attenuate inflammation, thus helping to prevent oral and neurological diseases.

2. **Microbiological interventions**: Treatments aimed at modifying the oral microbiome, such as the use of probiotics or specific antimicrobial agents, could prevent colonisation by **oral Streptococcus**. By re-establishing a healthy microbiological balance, these interventions can reduce the incidence of diseases associated with this bacterium.
3. **Vaccines**: The development of vaccines targeting the specific toxins produced by **Streptococcus buccale** represents a promising advance in the prevention of infections. These vaccines could induce an effective immune response, protecting the host from the harmful effects of the toxins.

Conclusion

This chapter has explored the clinical consequences of oxidative stress and toxins produced by **Streptococcus buccale**. The implications for oral and neurological health underline the importance of an integrative approach to the prevention and treatment of associated diseases. By taking account of the complex interactions between these factors, it is possible to develop more effective prevention and treatment strategies, thereby improving the quality of life of the individuals concerned.

Chapter 8: Oxidative stress defence mechanisms in Streptococcus buccale and other bacteria

Comparison of defence systems

Bacteria, like eukaryotes, have developed sophisticated mechanisms for coping with oxidative stress, a potentially destructive phenomenon caused by the accumulation of reactive oxygen species (ROS). These mechanisms include antioxidant enzymes, signalling molecules and specific metabolic pathways.

In **Streptococcus buccale**, the presence of superoxide dismutase (SOD) and catalase is essential for breaking down ROS generated by cellular metabolism and environmental factors, such as exposure to oxidising agents. SOD converts superoxide into hydrogen peroxide, which is then degraded by catalase. This detoxification capacity is comparable to that observed in human cells, illustrating an evolutionary convergence in the fight against oxidative stress.

In addition, other mechanisms, such as the production of antioxidant molecules and the modulation of signalling pathways, enable **Streptococcus buccale** to maintain its cellular integrity in the face of oxidative aggression. These strategies demonstrate the complexity of bacterial responses to oxidative stress and their ability to adapt to different environments.

Examples of other bacteria

- **Escherichia coli**

E. coli uses a range of antioxidant enzymes, including SOD and catalase, to neutralise peroxides. This bacterium is particularly robust in environments where oxygen levels can vary, making these enzymes crucial to its survival. In addition, **E. coli** can synthesise compounds such as glutathione, which helps protect against ROS and plays a role in the stress response.

- **Lactobacillus spp.**

Lactic acid bacteria, such as **Lactobacillus spp.** present in the human microbiota, exploit antioxidants such as lactic acid and glutathione to maintain the redox balance. Their ability to produce acid during fermentation helps to create an acidic environment unfavourable to pathogens, offering additional protection against oxidative stress and infection.

- **Pseudomonas aeruginosa**

Known for its virulence, **Pseudomonas aeruginosa** produces pigments such as pyocyanin, which have antioxidant properties. These pigments play a role in protecting against oxidative stress, while also contributing to the pathogenicity of this bacterium. **Pseudomonas aeruginosa** also uses quorum sensing mechanisms to modulate the production of its virulence factors in response to environmental conditions, thereby enhancing its survival and ability to cause infections.

Environmental and clinical impact

Oxidative stress defence mechanisms influence not only the survival of bacteria in hostile environments, but also their interaction with the host. For example, the ability of bacteria to neutralise oxidative stress can affect their virulence. Bacteria that manage to avoid or repair the damage caused by oxidative stress are often more pathogenic. This phenomenon is particularly worrying in the context of hospital-acquired infections, where antibiotic-resistant strains exploit these mechanisms to evade treatment.

In addition, these mechanisms open the way to potential therapeutic approaches. Targeting the antioxidant systems of bacteria could become a promising strategy for combating infections, notably through the development of drugs that inhibit essential antioxidant enzymes. By disrupting bacteria's ability to manage oxidative stress, it may be possible to increase their susceptibility to antimicrobial treatments.

Conclusion

This chapter has highlighted the importance of oxidative stress defence mechanisms in **Streptococcus buccale** and other bacteria. By understanding these systems, we can better understand their role in pathogenicity and resistance to treatments. This knowledge is essential for the development of new therapeutic strategies that could improve the management of oral infections and other associated pathologies.

Conclusion

This book has taken an in-depth look at the complex link between oxidative stress and the toxins produced by **Streptococcus buccale**, highlighting their impact on oral and neurological health. Through a series of chapters, we have addressed various aspects of this issue, highlighting several key points that are worth summarising.

1. **Interaction between oxidative stress and virulence**: We have shown that oxidative stress is a key factor in the pathogenesis **of Streptococcus buccale**. It not only influences toxin production, but also plays a crucial role in the bacterium's adaptation to a hostile oral environment. The mechanisms of detection and response to oxidative stress enable this bacterium to optimise its pathogenic capacities, thereby enhancing its virulence in a context where other micro-organisms might fail.
2. **Clinical implications**: The interactions between oxidative stress, toxins and oral and neurological diseases have significant clinical consequences. The presence of **Streptococcus buccale** in the oral cavity is linked to pathologies such as dental caries, gingivitis and even neurodegenerative disorders. This understanding opens the way to targeted therapeutic approaches, which could not only treat the symptoms but also tackle the underlying mechanisms of the associated diseases.

3. **Importance of future research**: Although significant advances have been made in our understanding of the mechanisms of action of **Streptococcus buccale**, much remains to be discovered. Further research is essential to elucidate the precise role of its toxins and oxidative stress in human pathology. In particular, an exploration of the interactions between **Streptococcus buccale**, the oral microbiome and the systemic environment could reveal critical connections for health.
4. **Multidisciplinary approach**: Finally, this study highlights the importance of a multidisciplinary approach, integrating microbiology, neurology and immunology. Such an approach is crucial to deepening our understanding of the effects of **Streptococcus buccale** on human health and to developing effective preventive and therapeutic solutions. By combining knowledge from different disciplines, we can better understand the complex interactions between microbiological factors and immune responses.

In sum, this book calls for a continued commitment to **oral Streptococcus** research, highlighting the importance of exploring these relationships to improve the prevention and treatment of oral infections and their systemic implications. We hope that this exploration will help to inform future investigations and stimulate innovations that could have a significant impact on public health.

Lexicon

- **Oxidative stress**: an imbalance between ROS and antioxidant systems.
- **Reactive oxygen species (ROS)**: Unstable molecules that can damage cells.
- **Gram-positive bacteria**: Bacteria that retain the violet dye during Gram staining.
- **Cellular respiration**: Process by which cells produce energy, often accompanied by the production of ROS.
- **Fermentation**: anaerobic metabolism of carbohydrates, also resulting in the generation of ROS.
- **Homeostasis**: A state of stable equilibrium in biological processes.
- **Oxidative stress**: A state of disturbed equilibrium between the production of reactive oxygen species (ROS) and the cell's antioxidant capacity, leading to cell damage.
- **Streptococcus buccale**: Gram-positive bacterium, common in human oral flora, known for its role in oral infections.
- **Superoxide dismutase (SOD)**: Enzyme which catalyses the conversion of superoxide into hydrogen peroxide, protecting cells from oxidative damage.

- **Hydrogen peroxide (H_2O_2)**: Intermediate product formed during the dismutation of superoxide, potentially toxic in high concentrations.
- **Glutathione peroxidase (GPx)**: Enzyme which uses glutathione to reduce peroxides, contributing to cellular protection.
- **Metallothioneins**: Proteins capable of binding metal ions and acting as antioxidants.
- **Oxidative stress:** a state of increased reactive oxygen species (ROS) in cells, leading to cell damage.
- **Streptococcus buccale**: Gram-positive bacterium present in the oral cavity, associated with various infections.
- **Blood-brain barrier (BBB)**: Protective barrier regulating the passage of substances between the blood and the central nervous system.
- **Neurotoxicity**: Toxicity with harmful effects on neurons and nervous tissue.
- **Inflammation**: Immune system reaction to aggression, often associated with pain and redness.
- **Vicious circle**: A situation in which a problem is made worse by the effects of a previous solution or condition.
- **Transcription factors**: Proteins that regulate gene expression by binding to specific DNA sequences.
- **Virulence**: The ability of a pathogen to cause disease.

- **Immune evasion**: mechanism by which a pathogen escapes recognition or destruction by the immune system.
- **Microbiome**: All the micro-organisms living in a specific environment, including the oral cavity.
- **Dental caries**: Damage to dental tissue caused by acids produced by bacteria.
- **Gingivitis**: Inflammation of the gums, often caused by a bacterial infection.
- **Periodontitis**: Serious inflammatory disease affecting the tissues supporting the teeth, which can lead to their loss.
- **Neurodegenerative**: Diseases leading to the degeneration and death of neurons.
- **Biofilms**: Communities of micro-organisms attached to surfaces, often protected by an extracellular matrix.
- **Reactive oxygen species (ROS)**: Unstable molecules derived from oxygen that can damage cells, proteins and DNA.
- **Superoxide dismutase (SOD)**: Enzyme which catalyses the conversion of superoxide into hydrogen peroxide.
- **Catalase**: Enzyme that breaks down hydrogen peroxide into water and oxygen.
- **Antioxidants** : Substances that neutralise free radicals and protect cells against oxidative stress.

- **Metabolism**: All the chemical reactions that take place in cells to maintain life.
- **Pathogenicity**: the ability of a micro-organism to cause disease.

References

1. Halliwell, B., & Gutteridge, J. M. C. (2015). Free Radicals in Biology and Medicine. Oxford University Press.
2. Marsh, P. D. (1994). Microbiology of dental plaque and its role in oral disease. *Journal of Clinical Periodontology*, 21(6), 411-430.
3. Rhee, S. G. (2011). Cell signaling. *H2O2, a key player in the oxidative stress response. Nature*, 475, 293-294.
4. Imlay, J. A. (2003). Pathways of oxidative damage. *Annual Review of Microbiology*, 57, 395-418.
5. Fridovich, I. (1997). Superoxide dismutases. *Annual Review of Biochemistry*, 66(1), 97-112. doi:10.1146/annurev.biochem.66.1.97
6. Halliwell, B., & Gutteridge, J. M. C. (2015). *Free Radicals in Biology and Medicine*. Oxford University Press.
7. O'Donnell, V. B., & Murphy, M. P. (2013). Antioxidants: A key to understanding the role of reactive species in health and disease. *Nature Reviews Molecular Cell Biology*, 14(3), 134-145. doi:10.1038/nrm3448
8. Imlay, J. A. (2003). Pathways of oxidative damage. *Annual Review of Microbiology*, 57, 395-418. doi:10.1146/annurev.micro.57.030502.090159

9. Berg, J. M., Tymoczko, J. L., & Stryer, L. (2002). *Biochemistry*. W.H. Freeman and Company.
10. Parnham, M. J., et al. (2014). "Oxidative Stress and Neurodegeneration." *Neuroscience & Biobehavioral Reviews*, 45, 345-360.
11. Wang, X., et al. (2019). "Mechanisms of Streptococcus-Induced Neuroinflammation." *Journal of Neuroinflammation*, 16(1), 100.
12. McEwen, B. S. (2008). "Stress, Adaptation, and Disease." *American Journal of Psychiatry*, 165(3), 331-343.
13. Watanabe, T., et al. (2020). "Pathogenic Mechanisms of Streptococcus buccale." *Frontiers in Microbiology*, 11, 1234.
14. Haghikia, A., et al. (2016). "The Role of Oxidative Stress in Neurodegeneration." *Nature Reviews Neuroscience*, 17(12), 797-810.
15. Fuchs, B. B., & Mylonakis, E. (2009). "The Role of Toxins in the Pathogenesis of Bacterial Infections." *Nature Reviews Microbiology*, 7(2), 103-114.
16. Chatzidimitriou, D., et al. (2020). "Oxidative Stress and Its Role in Bacterial Pathogenesis." *Frontiers in Microbiology*, 11, 237.
17. Nascimento, G. F., et al. (2018). "The Role of Bacterial Toxins in Immune Evasion." *Current Opinion in Microbiology*, 41, 56-61.

18. Pincus, S. H., et al. (2017). "Interplay Between Oxidative Stress and Bacterial Virulence." *Infection and Immunity*, 85(5), e00023-17.
19. Lichtenstein, S. J., & Regan, T. (2018). "Bacterial Stress Responses and Virulence." *Clinical Microbiology Reviews*, 31(3), e00022-17.
20. Haffajee, A. D., & Socransky, S. S. (2004). "Microbial Complexes in Subgingival Plaque." *Periodontology 2000*, 38(1), 36-58.
21. Kassebaum, N. J., et al. (2017). "Global, Regional, and National Prevalence and Years Lived with Disability for Oral Conditions." *Journal of Dental Research*, 96(4), 380-387.
22. Frazier, W. C., et al. (2019). "Oxidative Stress in Neurological Disorders." *Frontiers in Neurology*, 10, 1-9.
23. Dominy, S. S., et al. (2019). "Oral Infection: A New Risk Factor for Neurodegeneration." *Alzheimer's & Dementia*, 15(1), 124-133.
24. Khemiri, A., et al. (2021). "Antioxidants as Therapeutics Against Bacterial Infections." *Current Medicinal Chemistry*, 28(12), 2441-2461.
25. Halliwell, B., & Gutteridge, J. M. C. (2015). *Free Radicals in Biology and Medicine*. Oxford University Press.
26. Imlay, J. A. (2013). "The Molecular Mechanisms and Physiological Consequences of Oxidative Stress". *Nature Reviews Microbiology*, 11(9), 615-627.

27. Wasilenko, J., & Miller, R. S. (2014). "Role of Antioxidants in Streptococcus mutans Biofilm Formation". *Journal of Microbiology and Biotechnology*, 24(4), 502-510.

28. Slauch, J. M. (2011). "How the SoxRS System Works". *Annual Review of Microbiology*, 65, 223-240.

29. Fridovich, I. (1997). "Superoxide Dismutase". *Annual Review of Biochemistry*, 66, 97-112.

30. Cundliffe, E. (2000) "Antibiotic Production by Streptomyces". *Annual Review of Microbiology*, 54(1), 413-438.

MIX
Papier aus verantwortungsvollen Quellen
Paper from responsible sources
FSC® C105338

Printed by Books on Demand GmbH, Norderstedt / Germany